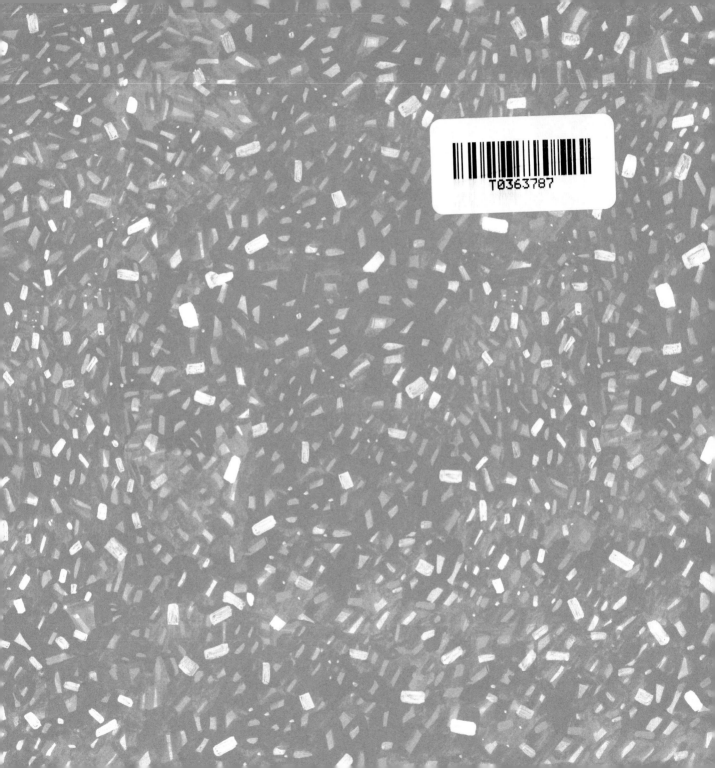

Crystals, Rocks, and Gemstones

EXPLORING EARTH'S TREASURES

Kelsey Oseid

TEN SPEED PRESS
California | New York

CONTENTS

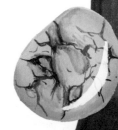

Crystals, Rocks, and Gemstones

For very many people a single
precious stone can provide a matchless
and perfect view of nature.

—Pliny the Elder, *Natural History*, first century CE

ROCKS, GEMS, CRYSTALS, AND MINERALS: AN OVERVIEW

Interest in minerals goes back as far as humans have been able to find them. From lumps of turquoise and spears of quartz to sheets of glittering mica and pieces of precious corundum, there are so many minerals that can catch our eye. Even without knowing the first thing about their origins, their names, or their chemical makeups, we find crystals simply fascinating. From the most traditional geologist to the most eccentric crystal mystic, no one can deny the powerful beauty of the perfect stone.

Few natural items are as treasured by collectors as minerals, crystals, and gems. Many people have an impulse to gather, hold, and keep these chunks of matter—even though, on some level, we know we can't really "own" them at all, since most of them have been around for billions of years and will outlast us by billions more. Putting a crystal on a shelf in your home is more akin to borrowing it than truly possessing it—except the lender, the Earth, will hardly notice the exchange.

Does the allure of a crystal really come just from its beauty? Sure, crystals are beautiful in the conventional sense. Amethyst, the purple variety of quartz, is an oft-cited favorite and is the cornerstone of many a crystal collection. It sparkles, shines, and has a lovely, gentle hue. But it isn't just surface-level beauty that draws us to crystals, since imitation crystals—made, say, out of cut glass or polished epoxy resin—lack the same appeal and are comparatively worthless. There's nothing rare or special about a fake plastic jewel formed by a human or a machine. It didn't develop its crystal form through millennia of hyperintense pressure in the mantle of the Earth, or explode from the center of a volcano before cooling in seawater. There's something that feels intrinsically powerful about a natural crystal's journey through time. We don't value just the look of a stone, but also its backstory—or are we talking about its life story, or its soul?

Of course, minerals are not living things. In fact, part of what defines minerals as a category is their inorganic origin. But we often can't help but personify them, ascribing living qualities to their beauty. When precious stones like diamonds are faceted to best enhance their appearance, we say they now have "more life"; in other instances, stones have different levels of "play" in the rapidly changing sparkles and glints as the stone is turned. Sometimes stones are named after body parts—for example, Pele's hair and Pele's tears are two natural artifacts that can result from volcanic eruptions, and gems with glinting inclusions are called "cat's eyes."

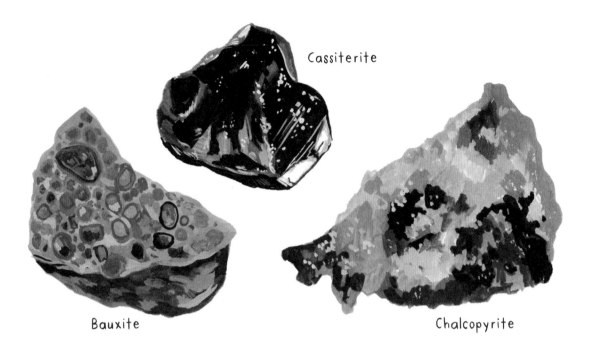

Cassiterite

Bauxite

Chalcopyrite

While minerals don't need us in the least—after all, humanity is barely a blip on the Earth's geological timeline—we certainly need them. Calcium is necessary for our teeth and bones; iron is essential for our blood; and potassium, sodium, and zinc have a multitude of health benefits. Minerals, whether obtained naturally from food sources or through supplements, are part of our daily lives. And, of course, so are the mineral rocks that comprise our buildings, roads, and other physical infrastructure. Then there's our technology—the phones, wearables, and other devices that are ever more embedded in the daily operations of our society. These devices wouldn't function without rocks and minerals like bauxite, the provider of gallium for LED displays; cassiterite, which comprises tin circuit boards; and chalcopyrite, the source of copper used for conducting heat and electricity. We are beholden to the power of minerals to keep our daily lives intact. This, combined with the mystique of crystals, the history revealed by cut and sculpted rock formations, and the beauty of gems, makes the study of rocks, gems, and minerals an essential part of understanding not just our Earth but also ourselves.

WHAT ARE ROCKS, CRYSTALS, AND GEMS?

Rocks are aggregates of minerals—inorganic materials, typically hard and with a crystalline structure. When minerals exist in large quantities in a relatively isolated state (as opposed to the small, jumbled mineral grains that tend to comprise rocks), they can form crystals. When rocks and crystals are cut or polished, they are called "gems"; minerals that are commonly used for the creation of gems are called "gem minerals." Typically made for decorative purposes, gems are most often found in jewelry and other decorative objects. Gems can also be created by refining any number of organic or human-made materials.

WHERE DO ROCKS, CRYSTALS, AND GEMS COME FROM?

The Earth is structured into three main layers: a central core surrounded by a thick mantle, which is surrounded by a thin layer called the "crust." Most gem minerals form within the crust or just below it in the upper mantle. In order for crystals to form, the right elements must be present alongside specific geological conditions, like heat and pressure. On the other hand, some materials we think of as gems come from organic sources, like ivory (animal tusks), amber (fossilized tree resin), and jet (fossilized wood).

Crust

Mantle

Outer Core

Inner Core

The "carat" used to measure crystals and minerals gets its name from the seeds of the carob tree.

WHAT MAKES A GEM VALUABLE?

Fine gemstones are graded on the "four Cs": color, clarity, cut, and carat. The color of a gem can make it more or less valuable, depending on the circumstances. In diamonds, a completely colorless diamond is typically the most valuable, but, in some cases, a particularly rare or beautiful color can increase a diamond's value. Gems can be idiochromatic, where their color is inherent to their chemical makeup—for example, peridot is always green. Or they can be allochromatic, which means they are variably colored by trace elements within a mineral, as in quartz (which can be purple, clear, black, pink, or orange, depending on which trace elements are present).

Clarity refers to how "clear" a stone is. Inclusions are any foreign matter, crystallizations, holes, or other "flaws" inside a crystal. The fewer inclusions, the higher the clarity. Higher clarity usually equates to higher worth, although some inclusions are sought-after, such as those present in dendritic agate, in which iron and manganese oxide form beautiful branches within transparent or translucent chalcedony.

Gemstones can be cut in nearly infinite styles. Certain styles of cut require more of the gemstone to be cut away, making these styles more expensive to produce.

A carat is a unit of weight, named after the carob tree, the seeds of which were once used as a standardized measure for weighing gems. Today, a carat is equivalent to 200 milligrams, a little heavier than an unpopped kernel of popcorn. Usually, the more a gem weighs, the more it is worth.

CRYSTAL MAGIC

Humans have long attributed certain spiritual powers or healing effects to various crystals, and many people continue to find crystals calming or otherwise generally beneficial. There is, of course, some tension between crystal mystics who believe in the healing properties of crystals and scientists who study rocks and minerals. But while these groups are unlikely to agree on whether crystals have special spiritual powers, most people of any or no spiritual persuasion can agree that the poignant beauty of crystals feels like a power all its own. As long as no one is hurt—say, by neglecting medical treatment in favor of a crystal cure—is there really any harm in believing in a little crystal magic here and there?

Crystal balls are used for scrying, a process by which a user gazes into a crystal in an attempt to access visions.

Crystal grids are arrangements of stones according to visual patterns. They can be done according to predetermined guidelines, or improvised in whatever shape is pleasing to the creator. Some say these arrangements help amplify the energetic powers of the component stones.

There are certainly many false narratives surrounding stones. These have tended to be unmasked over the centuries. For instance, in modern times we know that diamonds are not destroyed by goat's blood, even though people once believed they were. Some myths remain, however. For example, diamonds are not indestructible; though they are harder than other minerals, they can still crack, chip, or break. We can speak of the inherent value of gems and crystals as much as we like, but it seems the power of stones is usually ephemeral and subjective. Certain stones just speak to certain people. Here are some instances of stones that have captured the human spiritual imagination.

Hagstones are rocks with natural holes produced by erosion over time. They are said to bring good luck to their bearers, and some say fairies and spirits are visible through the hole in a hagstone.

Obelisks are often found in modern crystal shops. These carved towers, which can be made of virtually any common crystal, are said to focus the energy of a stone; they are also said to look lovely on a shelf.

JANUARY
Garnet

FEBRUARY
Amethyst

MARCH
Aquamarine

APRIL
Diamond

MAY
Emerald

JUNE
Pearl

JULY
Ruby

AUGUST
Peridot

SEPTEMBER
Sapphire

OCTOBER
Opal

NOVEMBER
Citrine

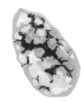

DECEMBER
Turquoise

BIRTHSTONES

The practice of assigning precious stones to different birth months, astrological signs, planets, and dates goes back thousands of years. One Vedic astrological text from 1300–1000 BCE associates different precious stones with the planets: emerald belongs to Mercury, yellow sapphires to Jupiter, blue sapphires to Saturn, and so on. Similarly, the ancient Romans associated different gems with astrological signs— these lists, like modern ones, vary in terms of which gems were associated with which birthdates. Over the centuries, such gemstone lists were co-opted by various groups, with Christians assigning gems to different apostles and Jewish writings attributing gems to different angels.

Nowadays, gem lists are determined by birthdate, usually assigning stones according to either birth month or astrological sign. The stones themselves vary too, depending on which list you're looking at. Here is one example of birthstone assignments commonly seen in the Western world: garnet for January, amethyst for February, aquamarine for March, diamond for April, emerald for May, pearl for June, ruby for July, peridot for August, sapphire for September, opal for October, citrine for November, and turquoise for December.

CRYSTAL SYSTEMS

The component cells of a given mineral crystal form a geometric framework; this is responsible for the characteristic shape of crystals. Every crystal can be classified into one of seven (or six, depending on who you ask) crystal "systems." These frameworks can each grow into a number of different forms. Astronauts aboard the International Space Station have been able to grow nearly perfect crystals, but on Earth, the near infinite complicating environmental factors, including gravity and constraints from the surrounding rocks in which a crystal grows, create a huge variety of forms, often even more beautiful and intriguing than a theoretically perfect crystalline shape.

Isometric or cubic crystal systems are common. They have three axes of equal length that intersect at right angles. The shapes they can form include cubes and octahedrons.

Tetragonal systems are uncommon. They have three axes perpendicular to one another, with two of equal length. They can form tall and short prisms.

Hexagonal systems have three equal-length horizontal axes crossing each other at 60 degrees, with a fourth perpendicular axis. Trigonal systems are sometimes classified as a subsystem of hexagonal systems. Hexagonal systems often form hexagonal prisms.

Orthorhombic systems have three unequal axes at right angles to one another. They often form tabular or prismatic shapes.

Monoclinic systems have three unequal axes with just two crossing at right angles. They often form tabular shapes.

Triclinic systems are the least symmetrical of them all, with three axes of unequal length and none perpendicular to each other.

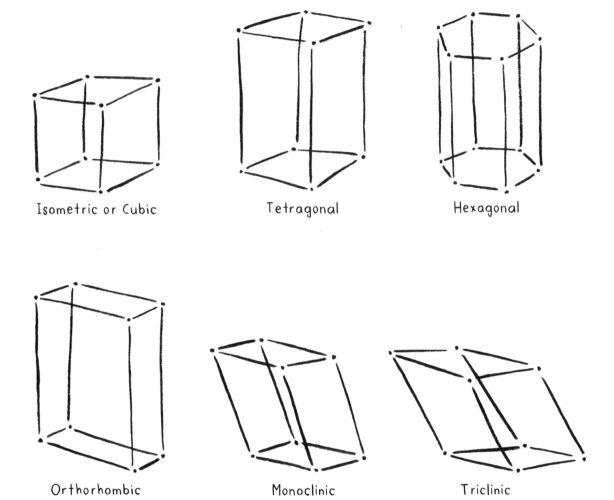

Isometric or Cubic

Tetragonal

Hexagonal

Orthorhombic

Monoclinic

Triclinic

CRYSTAL HABITS

A crystal's habit is the characteristic shape of its form. Acicular crystals are needle-like. Arborescent crystals occur in branching lusters. Bladed crystals are flattened and long, like a knife blade. Botryoidal crystals are blob-like, with a surface covered in clusters of sphere-like shapes. Capillary crystals, also known as "filiform crystals," are very fine or hair-like. Columnar crystals are composed of parallel columns. Dendritic crystals are similar to arborescent crystals, but finer. Drusy crystals are identified by a coating of small crystals on the surface of another crystal or rock. Foliated crystals, also called "micaceous crystals," occur as thin, layered sheets.

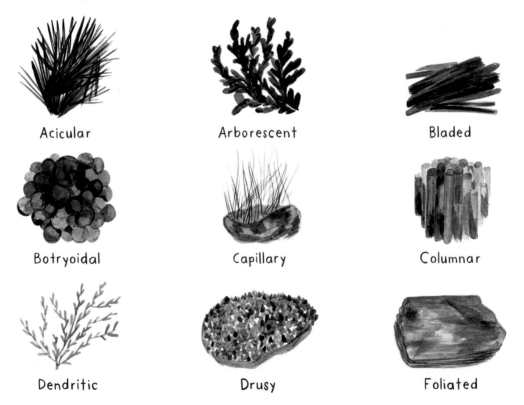

Acicular

Arborescent

Bladed

Botryoidal

Capillary

Columnar

Dendritic

Drusy

Foliated

Globular crystals are similar to botryoidal crystals, but with fewer and larger blobs. Massive crystals have no distinctive crystalline structure, appearing instead as solid masses. Oolitic crystals occur as bunches of tiny, fish-egg–like spheres. Radial crystals grow outward from a central point. Reniform crystals are blobby and kidney-like. Reticulated crystals grow as net-like lattices of long, slender crystals. Rutilated crystals show needle-like inclusions within them. Striated crystals have parallel grooves running along their surfaces. And tabular crystals are tablet- (or book-, or plate-) shaped.

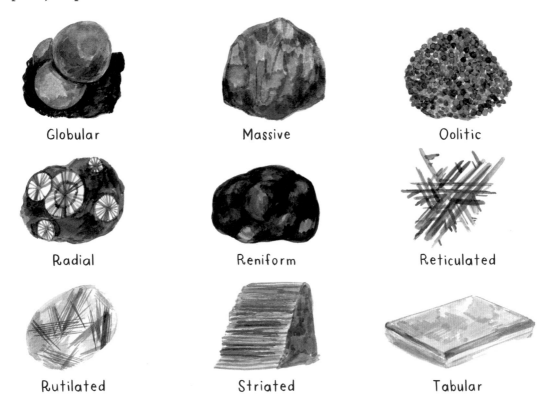

Globular   Massive   Oolitic

Radial   Reniform   Reticulated

Rutilated   Striated   Tabular

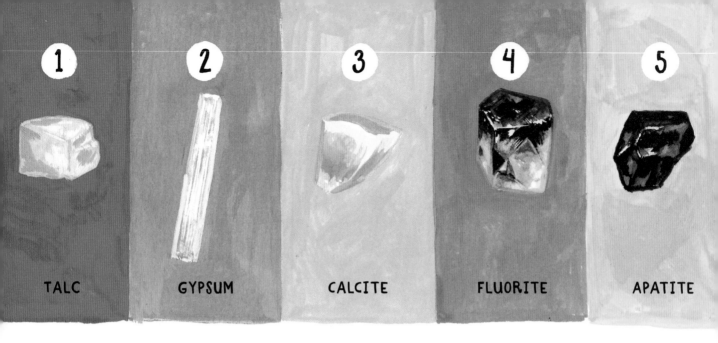

1	2	3	4	5
TALC	GYPSUM	CALCITE	FLUORITE	APATITE

HARDNESS

Comparing the hardness of different stones against one another has long been a key component of gem identification. The Mohs scale, named after mineralogist Friedrich Mohs, is the most commonly used measurement of hardness today. The scale ranges from 1 to 10, with 1 representing the softest possible mineral (talc) and 10 the hardest (diamond). Although the harder of two minerals can scratch the softer, hardness does not equate to strength! Hard minerals can still be brittle and breakable.

Everyday objects can also be reliably placed along the scale to provide a quick point of reference. A fingernail measures around a 2.5, a penny a 3.5, a glass plate a 5.5, a steel nail a 6.5, and a masonry drill bit an 8.5.

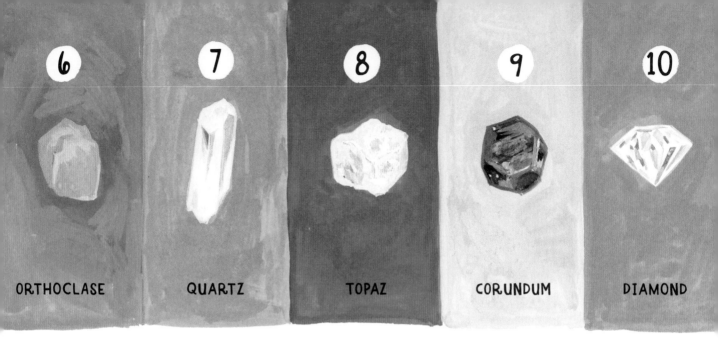

6	7	8	9	10
ORTHOCLASE	QUARTZ	TOPAZ	CORUNDUM	DIAMOND

OTHER IDENTIFYING FACTORS

In addition to observations of habit, hardness, and color, crystal identification relies on factors such as specific gravity, streak, cleavage, and fracture. Specific gravity is a measure of density, which essentially determines how "hefty" a given stone is. Streak is the color left behind by a stone when it's rubbed against a piece of unglazed porcelain, known as a "streak plate," a common item in geology tool kits. Many stones streak white, but some streak red, pink, green, blue, black, or colorless. A stone's color does not necessarily determine the color of its streak—for instance, pyrite streaks black despite its brassy, golden surface. Cleavage and fracture refer to different ways in which minerals can break. If a stone cracks cleanly along one or more planes, that's referred to as "cleavage." If it shatters with no clear delineation of planes, that's called "fracture."

CRYSTAL ODDITIES

Some crystals and gems appear to be different colors, depending on the angle from which they are viewed. This prized effect is called "pleochroism" and is commonly found in andalusite, which shifts from green to red; benitoite, which shifts from colorless to blue; and thulite, which shimmers yellow or pink, as well as many others.

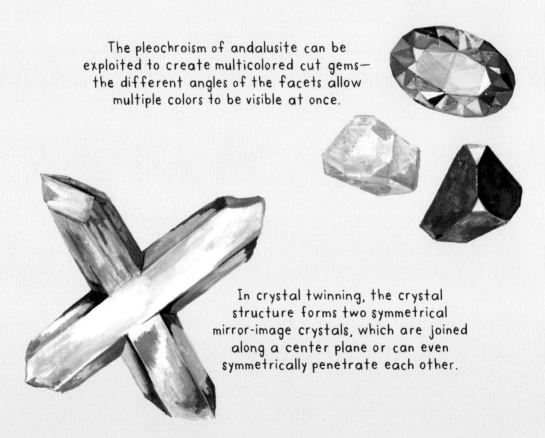

The pleochroism of andalusite can be exploited to create multicolored cut gems— the different angles of the facets allow multiple colors to be visible at once.

In crystal twinning, the crystal structure forms two symmetrical mirror-image crystals, which are joined along a center plane or can even symmetrically penetrate each other.

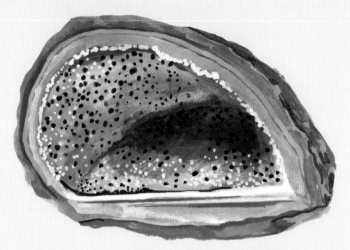

Geodes are hollow cavities lined with crystals.
Geodes form when minerals seep into these
hollows; over time, the mass around the vaguely
spherical cavity erodes, leaving just a shell
of the original material, still round but now
also filled with hidden treasure.

Chatoyancy, also called the "cat's-eye
effect," is an optical effect in which
a band or bands of light seem to shimmer
just below the surface of a polished
gemstone. This is a result of parallel
inclusions or fibers within the stone.
Tiger's eye, a metamorphic rock displaying
a chatoyant effect, is a commonly
found example in crystal shops.

FLUORESCENCE

Some crystals are fluorescent, emitting a variety of colors under UV light. This characteristic is inconsistent; different specimens of the same mineral type can fluoresce in different colors or none at all. The name comes from fluorite, which often, but not always, fluoresces (when it does, it usually fluoresces blue). Fluorescence is relatively rare among minerals and is prized for its otherworldly strangeness and beauty. Some crystals are even phosphorescent—capable of storing and emitting an afterglow after the UV light source is removed.

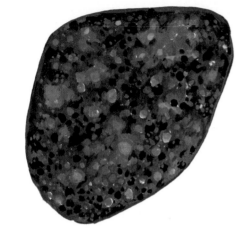

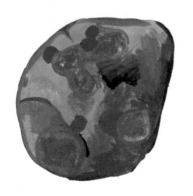

The Murchison Snuff Box was an 1867 gift from Tsar Alexander II of Russia to geologist Sir Roderick Murchison. The sixteen diamonds set into its lid, all of which appear clear in regular light, reveal very subtle differences—caused by different quantities of various trace elements in their makeup—when they fluoresce in different colors.

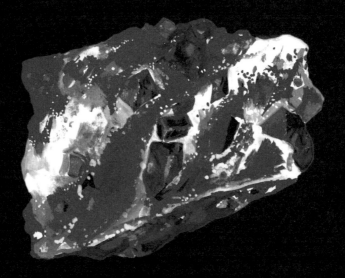

Recently, with the much wider availability of tools such as UV flashlights, more fluorescent crystals have been discovered. At least one special new fluorescent crystal has been identified by UV light-wielding beachgoers: the stone, nicknamed "Yooperlite," was found on the stony beaches of Lake Superior in the Upper Peninsula of Michigan.

FAMOUS MINERAL COLLECTIONS

The exterior walls of Chicago's Tribune Tower are embedded with close to 150 samples of historically significant and scientifically interesting stones. These samples were added throughout the past century and include a lump of rock from the dome of St. Peter's Basilica in Rome, a piece of the Berlin Wall, a stone head from London's House of Parliament, a fragment of the Great Wall of China, and a figure from a sixth-century Chinese temple. The stones are not behind a glass barrier or protected from touch in any way, save for one: a moon rock bequeathed by legendary astronaut Buzz Aldrin. The glass-protected moon rock remained on the tower for several years before its removal, and was returned to National Aeronautics and Space Administration (NASA) at their request in 2011. There's still discussion about replacing it with another space rock, perhaps this time one that's out from behind the glass.

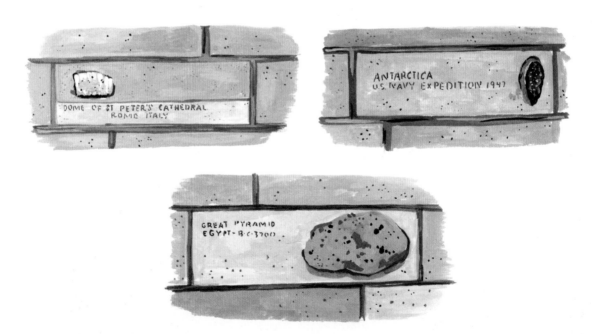

The Smithsonian's National Museum of Natural History houses one of the largest mineral and gem collections in the world, with around 350,000 individual mineral specimens and 10,000 gems. The start of the collection—and of the museum itself—began with a gift of 8,000 mineral specimens from James Smithson, after whom the Smithsonian is named. The collection houses the infamous Hope Diamond (about 45 carats), along with other famous gems like the Mackay Emerald (over 167 carats), the Carmen Lúcia Ruby (over 23 carats), and the Whitney Flame Topaz (over 48 carats), each among the largest of their kinds. The Smithsonian's Hall of Geology, Gems, and Minerals is, like the rest of the National Museum's exhibits, free to visit, allowing the treasures once held in private collections behind closed doors to be enjoyed by all.

There's even a mineral named after Smithson: smithsonite, a zinc carbonate that can be many different colors but is most prized in bright green or blue.

CUTTING AND TREATING GEMS

A cut and polished stone is generally considered more valuable than the raw material it came from, especially now that today's gem-cutting technology and equipment are more sophisticated than ever before. Historically, precious stones were often left in a relatively unprocessed state, due to the belief that the power of the stone came more from its weight than from its brilliance. Many of history's largest precious stones have now been recut into smaller, more symmetrical gems. The largest diamond ever found, the Cullinan Diamond, which was over 3,100 carats when extracted from a mine in South Africa, was chopped into nine large-faceted stones and nearly one hundred smaller ones.

Even when stone size was most valued, manipulation of stones, minerals, and other natural materials has been a part of human culture for thousands of years. The earliest artisans pierced holes through stones and other materials to create beads. Later, a technique called "bruting" emerged, in which softer materials were rubbed against harder ones in order to shape them. Certain materials on the softer end of the spectrum, like turquoise, amber, and lapis lazuli, are easier to carve and have long been used for more sculptural lapidary work, but the carving of harder materials like jade and amethyst was also mastered centuries ago. The drive to create something beautiful out of precious materials is strong.

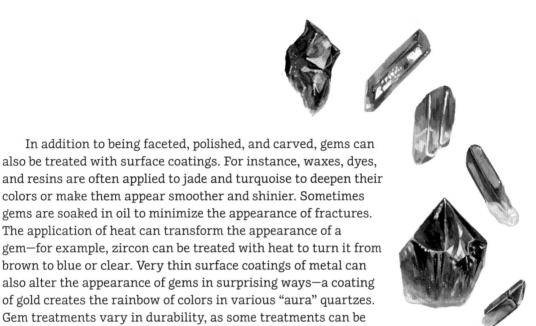

In addition to being faceted, polished, and carved, gems can also be treated with surface coatings. For instance, waxes, dyes, and resins are often applied to jade and turquoise to deepen their colors or make them appear smoother and shinier. Sometimes gems are soaked in oil to minimize the appearance of fractures. The application of heat can transform the appearance of a gem—for example, zircon can be treated with heat to turn it from brown to blue or clear. Very thin surface coatings of metal can also alter the appearance of gems in surprising ways—a coating of gold creates the rainbow of colors in various "aura" quartzes. Gem treatments vary in durability, as some treatments can be scratched or chipped off quite easily.

Preciousness

The hardest gems, like diamonds, rubies, sapphires, and emeralds, are sometimes called "precious stones," while softer minerals, such as garnet and turquoise, are referred to as "semiprecious stones." These categories are not very clearly defined and are therefore not generally useful, since a precious stone isn't necessarily more expensive than a semiprecious stone. Additionally, we often use the phrase precious stone colloquially to refer to any sought-after mineral. With so many beautiful crystals in the world, why make it a contest?

Round cut

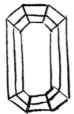

Emerald cut

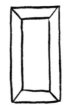 

Baguette cut

Pear cut

Asscher cut

Marquise cut

Cushion cut

Princess cut

Scissor cut

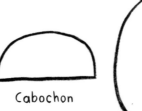 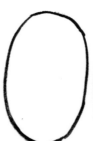 

Cabochon

Gem Cuts

Cutting, polishing, engraving, or otherwise shaping a mineral is what turns it into a gem. Gemstones are cut in a variety of ways, often aimed at bringing out a given stone's best qualities, and sometimes simply to fit into a jewelry setting; some gemstones are even carved as sculpture themselves.

The *cabochon* is a simple, polished form of gemstone, usually convex with a relatively flat back. This is an older cutting style that is still commonly used today. Cabochons are popular for stones with chatoyancy or the related, star-shaped "asterism" effect, as they can maximize their visual impact. Some are "double cabochons" and are domed on both sides.

The process of creating faceted gemstones, usually cut to maximize sparkle, has its roots in early human history, when Stone Age people smashed harder rocks against softer ones to shape them into the earliest tools. That same principle was at the root of the earliest gem cutting—rubbing harder gems against softer ones to shape them. Gem cutting has improved over time as people mastered cutting and polishing crystals according to their internal structure and optical properties; today, lasers are used for precision shaping and cutting. Faceting can manipulate the fire (the dispersion of white light into a spectrum of color), scintillation (the flashes of light seen when a gem is moved), and brilliance (the overall brightness of a gem when viewed from above) of a given stone.

A 20-sided dice, or "D-20," used to play tabletop and role-playing games like Dungeons & Dragons, can be cut from crystals like amethyst.

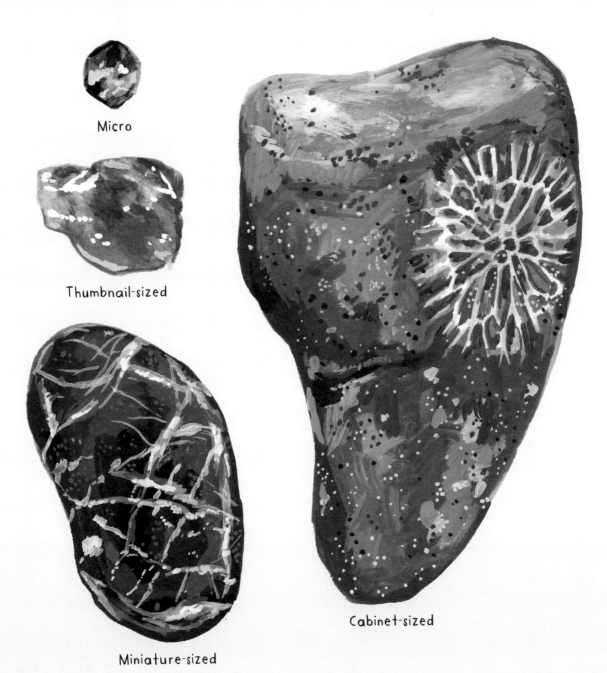

Micro

Thumbnail-sized

Miniature-sized

Cabinet-sized

SPECIMEN SIZES

The size of a mineral specimen is usually given in inches or centimeters. Specimen size is also described by a plethora of umbrella terms based on general measurements. The actual limit on these categories varies somewhat depending on the source, but, generally speaking, samples under a centimeter (or around half an inch) in any direction are "micro-sized," samples larger than micro but under 2.5 centimeters (around an inch) in any direction are "thumbnail-sized," those larger than thumbnail but under 5 centimeters (around two inches) are "miniature-sized," and anything larger than miniature is "cabinet-sized." Cabinet specimens can also be grouped into small, medium, and large cabinet categories, with museum-cabinet specimens measuring up to several feet or more.

In order to protect them and to aid in microscopic observation, the smallest specimens are commonly affixed inside a box known as a "micromount."

MINERAL SOURCES FROM AROUND THE WORLD

Most gem-quality minerals form in the Earth's crust and can be found on every continent. Different minerals are concentrated in different places, and as new deposits are found and old sources are depleted, the primary source of a given mineral can shift. India, for example, was an early source of diamonds; after that supply diminished via mining, Brazil, and then South Africa, emerged as new sources of diamonds. Some rare minerals are only found at a single site, such as tanzanite (discovered only in one location in Tanzania) and benitoite (found only in California).

 Agate
 Amethyst
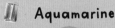 Aquamarine
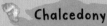 Chalcedony
 Coral
 Diamond
 Emerald
 Garnet
 Jadeite
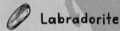 Labradorite

Nephrite
Peridot
Quartz
Ruby
Sapphire
Sodalite
Topaz
Tourmaline
Turquoise

 Agate
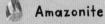 Amazonite
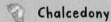 Chalcedony
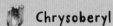 Chrysoberyl
 Diamond
 Emerald
 Garnet
Lapis Lazuli

Opal
Pearl
Quartz
Rhodochrosite
Sodalite
Topaz
Tourmaline

Agate
Amber
Fluorite
Quartz

Agate
Amethyst
Aquamarine
Chrysoberyl
Diamond
Emerald
Garnet
Rhodochrosite
Ruby
Sapphire
Spinel
Tanzanite
Topaz
Tourmaline

Alexandrite
Amber
Amethyst
Aquamarine
Chalcedony
Chrysoberyl
Coral
Diamond
Emerald
Garnet
Jadeite
Lapis Lazuli
Malachite

Moonstone
Nephrite
Pearl
Peridot
Quartz
Ruby
Sapphire
Sodalite
Spinel
Topaz
Tourmaline
Turquoise
Zircon

Coral
Diamond
Emerald
Nephrite
Opal
Pearl
Sapphire

ROCKHOUNDING

Removing and collecting rocks, minerals, and fossils from nature is called "rockhounding." It's an excellent way to interact with nature—it can be free, or very low cost if you invest in basic supplies like a magnifying glass, hammer, and chisel—and it's possible to get started with very little knowledge of rocks. Just going to the beach and sifting through pebbles to find ones that feel special to you is a type of rockhounding.

Advanced rockhounds often search for specific stones using more sophisticated approaches. For instance, one might look for "companion stones"—less valuable stones that occur in abundance among more valuable ones—because they can give a clue as to where those more valuable stones might be found. For agate hunters, this might mean looking for gravel beds, which commonly accompany agate deposits. You might even keep an eye out for where stones have been moved by people—like extra gravel removed from a quarry—since that can indicate the presence of more valuable or interesting finds, even in human-made spaces.

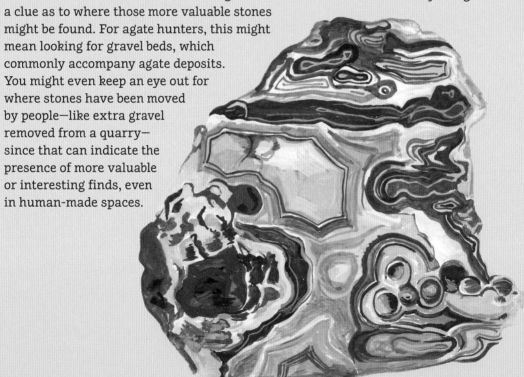

Collecting rocks responsibly is important. Every area has its own regulations regarding rockhounding, and it's a good idea to check posted signage and even to consult with local authorities to make sure you're following any rules and being respectful of the site from which you're collecting. In the United States, rockhounding is allowed on national forest land, but may require a permit—and certain items are usually off-limits for collection, like meteorites and cultural artifacts. Casual collection of abundant fossils is usually allowed, and since much of prehistoric North America was underwater, fossilized organisms are more abundant in our forests than one might think. Find out what's in your area!

Generally speaking, it's a good idea to follow the same best practices you would any time you're out in nature. Keep only a small quantity of what you find and leave the rest for others to enjoy. Tread lightly and disturb wildlife as little as possible. Certainly don't leave litter behind, and bring a bag to collect any litter you do encounter. Any time you spend out in nature is an opportunity to leave it cleaner than you found it. Steer clear, of course, of sacred sites and burial grounds. (If you're not sure if you could be near or on a sacred site, now is a great time to do some research and learn more about the area you're planning to search.)

ETHICAL CONCERNS

Cheap gems and crystals are more readily available to the average person than ever before, but it's important to remember that low prices don't always equate to low cost of labor. Many of the crystals available for sale in person and online are mined in unsafe or unethical conditions. When acquiring a crystal for its purported healing properties or energies or for scientific study, we must look deeper into the crystal's backstory and consider the lives of those who brought it to us. Surely if mystical energies are real, then these stones carry some memory of their pathways to our display shelves.

The same rule applies to organically sourced treasures, perhaps even more so. Coral can be unethically or illegally sourced under conditions that are not merely potentially unsafe, but also endanger the already threatened coral reefs of the world. Demand for elephant ivory has famously led to a dangerous culture of poaching, endangering this iconic animal. Conservation organizations have put in immense effort to create international initiatives aimed at curbing demand for elephant ivory; although these initiatives have helped reduce poaching, the biggest potential threat to elephants now comes from illegal demand. Elephant ivory should never be purchased. It is important to note, though, that the legal trade of walrus ivory is culturally and economically important for Alaska Native peoples and has not had the same negative impacts as that of the illegal elephant ivory trade.

As with any precious material, the more you can find out about its source before acquiring it, the better. And remember, too, that many museums, including those with public and free admission collections, have gems and precious materials on display for all of us to see, learn from, and admire. Perhaps a sustainable and ethical model of gem collecting is one rooted in collectivism rather than individual ownership.

MINERALS

To be considered a mineral, a substance must be naturally occurring, relatively homogenous, and solid, with a well-defined chemical composition and a crystal structure, even at a very small scale. Mineral "species" with different crystal structures can be distinct from one another even if they share the same chemical composition. Over five thousand different kinds of minerals are currently recognized. They are classified based on their chemical compositions. These categories include borates, carbonates, halides, silicates, sulfates, sulfides, oxides, phosphates, and native elements (minerals composed of atoms from a single element, like gold, sulfur, or silver).

Several different processes can take place to form crystals large enough for humans to see, feel, and hold. Some mineral crystals form during the cooling of hot magma. Others form as the chemicals in mineral-rich water turn solid after evaporation. And mineral crystals can also transform from one type to another based on their environment, due to oxidation, rusting, or recrystallization resulting from heat and pressure.

There's no doubt that crystals are some of the most beautiful naturally occurring objects on our planet, and it's no wonder that humans have long imbued them with mystical powers and great value. Here is just a sampling of some of the most beautiful minerals known to humankind.

FLUORITE

Class: Halides / Hardness: 4

Fluorite can form during the cooling of hydrothermal solutions, solidifying into a "vein" of crystals in the Earth. Moderately soft, it is the standard for a 4 on the Mohs scale. It can occur in a whole rainbow of colors, perhaps more than any other common crystal. Because it is brittle and fractures easily, it is not often cut or turned into jewelry for daily wear; nevertheless, because of its beauty, it has been carved into collectible objects for thousands of years; "murrina vasa," a type of luxury tableware carved from fluorite, was a Roman status symbol.

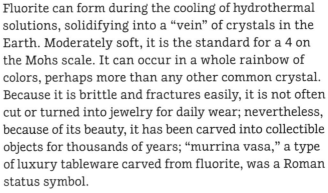

CALCITE

Class: Carbonates / Hardness: 3

Among the most common minerals on Earth, calcite can occur in several notable forms. The long crystals of "dogtooth spar" give it its fitting name. There's also "nailhead spar," which looks like a cluster of crystal nails. The remarkable transparency and other optical properties of "Iceland spar" has led to its use in specialized scientific equipment, like polarizing microscopes. Calcite is the standard for a 3 on the Mohs scale.

Dogtooth spar

Iceland spar

Nailhead spar

PYRITE

Class: Sulfides / Hardness: 6–6.5

The name *pyrite* comes from the Greek word *pyr*, meaning "fire," since pyrite sparks when struck with iron. Pyrite can naturally form into a near-perfect cube, giving it a human-made appearance. It can also be present alongside other minerals and rocks—the golden flecks found in lapis lazuli are actually pyrite!

Occasionally, pyrite found in coal mines can take the form of flat, radial discs known as "miner's dollars." These mineral aggregates are also referred to as pyrite "suns."

GALENA
Class: Sulfides / Hardness: 2.5

Galena, a lead ore (a material from which metal can be extracted), has long been the main source of lead. Lead contamination, even in tiny doses, can cause extreme negative health consequences, but galena is generally considered safe to handle and is sold in crystal shops (keep it away from children and pets, though, since it is still unsafe to ingest). Galena can also be mined for its silver content.

STIBNITE
Class: Sulfides / Hardness: 2

Stibnite often forms long, striated crystals, and is mined as an antimony ore. Antimony was traditionally used to make kohl, a cosmetic used to line the eyes; modern kohl is often made from galena (above) instead. Kohl is banned in many countries, including the United States, because of the health consequences of its heavy metal content.

CORUNDUM

Class: Oxides / Hardness: 9

Corundum is an aluminum oxide with a hardness nearly on par with diamond. Pure corundum is colorless, but minor mineral impurities can create a range of colors. Iron and titanium, for instance, are responsible for the blues and yellows of the precious gemstone sapphire, and chromium creates the red of rubies.

Some sapphire and ruby corundum can have inclusions of rutile strands, which create an optical phenomenon called an "asterism." Once polished into cabochons, these gemstones are referred to as "star sapphires" and "star rubies."

Asterisms in star sapphires and rubies

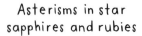

CORUNDUM: Ruby
Class: Oxides / Hardness: 9

The name *ruby* once referred to a number of different red gemstones, but today, only the deep red variety of corundum is considered a true ruby.

CORUNDUM: Sapphire
Class: Oxides / Hardness: 9

Sapphires are simply colorful variants of corundum and come in green, yellow, purple, orange, and more colors. Padparadscha sapphires, named after the Sinhalese for "lotus-color," are sapphires with a peachy pink hue.

Padparadscha
sapphires

SPINEL

Class: Oxides / Hardness: 8

Like corundum, spinel can have rutile strand asterisms. Spinel can occur in a variety of colors from blood red to blue and green. Synthetic spinel is a popular cut gem in modern jewelry.

SPINEL: **Magnetite**

Class: Oxides / Hardness: 5.5–6.5

The discovery of magnetism by ancient humans is credited to magnetite. Magnetic magnetite, referred to in ancient times as "lodestone," was used to magnetize the iron of the very first compasses. What magic!

CHRYSOBERYL

Class: Oxides / Hardness: 8.5

Chrysoberyl is another particularly hard gemstone. The variety known as "alexandrite" appears to change color depending on the light in which it is viewed. Under indoor incandescent lighting, alexandrites are pink, whereas when illuminated by natural daylight, they appear blue-green.

Alexandrite under
incandescent lighting

Alexandrite in daylight

HEMATITE

Class: Oxides / Hardness: 5.5–6.5

Hematite is an iron oxide named after the ancient Greek word for "blood." Hematite was once thought to promote liver and bladder health, and can occur in a formation called "iron rose"—a round form composed of layered flat crystals. Iron oxides such as hematite contribute various shades of red, orange, and yellow to ocher and sienna pigments. Ocher pigments have been used for hundreds of thousands of years. In fact, there is evidence that these pigments, as well as manganese pigments, were made in East Africa as far back as 320,000 years ago! Similar pigments were used in the famous Lascaux caves in France. When heat-treated, ocher and sienna can create a wide range of different warm tones that have been popular with artists since ancient times, such as yellow ocher, red ocher, raw sienna, and burnt sienna.

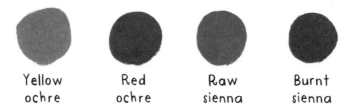

Yellow ochre   Red ochre   Raw sienna   Burnt sienna

CHRYSOCOLLA

Class: Silicates / Hardness: 2–4

Although chrysocolla is sometimes used as a gemstone, it has historically been more significant as a source of pigment. Chrysocolla is responsible for the blue-green color found in paints decorating the walls of ancient Egyptian tombs.

REALGAR

Class: Sulfides / Hardness: 1.5–2

Another source of toxic pigment, realgar is named after the Arabic *rahj-al-ghar*, meaning "powder of the mine." Realgar was used as an orange paint pigment until its displacement in the eighteenth century, when less toxic alternatives were developed.

CINNABAR

Class: Sulfides / Hardness: 2–2.5

Cinnabar ranges in color from bright to dark red and is a common source of mercury.

ORPIMENT

Class: Sulfides / Hardness: 1.5–2

Orpiment's name comes from *auripigmentum*, the Latin for "gold paint." As its name suggests, it was once used as a yellow pigment in paint, but its usage declined because of its highly toxic arsenic composition.

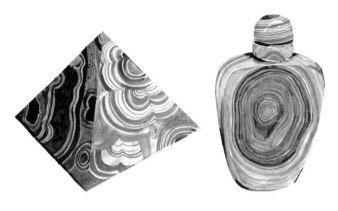

MALACHITE

Class: Carbonates / Hardness: 3.5–4

Malachite can have a lovely, bubbly botryoidal form that produces an agate-like "bull's-eye" pattern when sliced crosswise. In ancient Egypt, ground malachite was used as a cosmetic; later, it was used to paint Italian frescoes.

AZURITE

Class: Carbonates / Hardness: 3.5–4

During the Middle Ages, azurite was used as a pigment in illuminated manuscripts. It was a popular alternative to lapis lazuli for a time, but the blue pigment ultimately proved to be unstable because powdered azurite turns into malachite when exposed to moisture. Consequently, some paintings originating from this time are now green when they were originally blue.

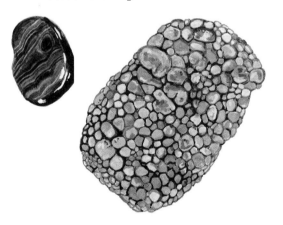

A slice of rhodochrosite "bacon"

RHODOCHROSITE

Class: Carbonates / Hardness: 4

Often called "Inca rose," rhodochrosite is the national stone of Argentina, where it has been found in abandoned Inca mines as stalactites, the icicle-shaped formations that hang from cave ceilings. Raspberry-pink rhodochrosite, whose color comes from its manganese content, can form as either a transparent crystal or a distinctively banded stone, which, incidentally, looks remarkably similar to bacon when thinly sliced.

A cross section of a rhodochrosite stalactite reveals bright concentric banding.

TURQUOISE

Class: Phosphates / Hardness: 5–6

Turquoise's name likely comes from Türkiye (Turkey), where the first turquoise was sourced and exported to Western Europe. Once believed to have the power to cure cataracts, turquoise has been worked as a gem for thousands of years, and was used in ancient Mesopotamian jewelry.

The gold mask of Tutankhamun bears two animal figures—the vulture, representing Upper Egypt, and the cobra, representing Lower Egypt, the latter of which is decorated with lapis lazuli, turquoise, and carnelian. The god of the dead, Osiris, was said to wear a turquoise crown.

The mask of
Tutankhamun

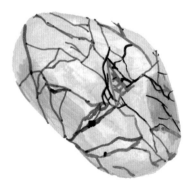

HOWLITE

Class: Borates / Hardness: 3.5

Howlite is most notable for its use as a turquoise substitute—it is often dyed blue and sold as turquoise to unsuspecting buyers. It can be distinguished from true turquoise by by its lower hardness; sometimes it is sold undyed as "white turquoise." Howlite is typically found as massive nodules, and can be tumbled, cut, and carved.

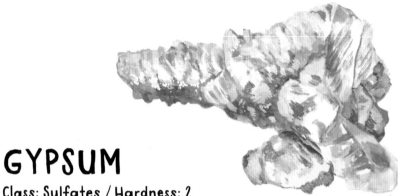

GYPSUM

Class: Sulfates / Hardness: 2

Gypsum naturally occurs as either twinned crystals or long,
translucent spars, known as "selenite." The wand form of ever-
popular selenite is a result of the mineral splitting lengthwise,
a phenomenon that's similar in appearance to splitting wood.
Selenite wands are popular among modern-day crystal healers.
They are said to "cleanse" everything from one's personal energy
to physical objects and even other crystals. As is typical with
healing crystals, it seems the power of selenite wands comes
from their beautiful form, as the gypsum they're composed of is
also the main component of plaster and drywall, which generally
have less mystical power ascribed to them. Gypsum can also
form a conglomeration of flat crystal plates grouped together in
a rosette known as a "desert rose."

Desert rose gypsum

Selenite "wand"

CELESTINE

Class: Sulfates / Hardness: 3–3.5

Celestine, also known as "celestite," can be colorless, yellow, or even red, but its most popular form is a soft sky blue from which it takes its name (after the Latin *coelestis*, for "celestial" or "heavenly"). It can form geodes, and sometimes even forms within empty cavities inside of fossilized ammonites. It's a primary source of the element strontium, which gives color to fireworks (though, in the enigmatic ways of mineral alchemy, the color it provides is not blue but red).

Barium is also used in barium x-rays, diagnostic procedures in which patients swallow barium powder to coat the inside of the gastrointestinal tract. Barium appears white on x-ray film, so images taken during this procedure illuminate the GI tract and can help doctors diagnose GI-related health issues.

Desert rose barium

BARITE

Class: Sulfates / Hardness: 3–3.5

Like gypsum, the sulfate mineral barite (sometimes spelled *baryte*) can also aggregate into the rose-like formation called a "desert rose," often with sand granules as inclusions. These can even take on a red hue, making them look very floral indeed. Barite is the main source of the element barium, which is used in fireworks (like celestine)—it imparts a green color. In fact, barite and celestine can be confused for each other, particularly when a celestine specimen doesn't have its characteristic blue color; burning small amounts of the ground-up mineral can help distinguish the two based on the color of the resulting flame.

DIAMOND

Class: Native elements / Hardness: 10

The hardest natural material known to humanity, diamonds are also perhaps the most coveted. For much of human history, diamonds were mined only in India, where they were thought to ward off dangers of all kinds, from the physical to the spiritual.

Diamonds are graded more intensely than any other gem. Mineral impurities can create diamonds that are brown, yellow, blue, green, red, pink, or even black (known as "carbonado" diamonds), and these colorful diamonds are considered to be more valuable than those that are clear. Stones with better clarity are generally valued more highly than those with inclusions or flaws. Diamond cutting is typically focused on maximizing the stone's interaction with light: the brighter and more sparkly the results, the more expensive the diamond.

Diamonds are actually relatively common among precious gems. Diamond engagement rings were popularized by an ad campaign funded by the De Beers mining company, which sought to artificially inflate the value of diamonds and monopolize the diamond industry. Their 1947 "A Diamond Is Forever" ad campaign famously cemented the conventional belief that diamonds are wedding ring essentials. Concerns over "blood diamonds"—those mined in conflict zones or under unsafe conditions, or used to fund wars—have led some to seek out alternatives, such as lab-grown diamonds, other stones, or alternative jewelry options.

The Hope Diamond in
its current setting

The Koh-i-Noor

The Daria-i-Noor

Famous Diamonds

The largest and most valuable diamonds in history are also
the most infamous. Perhaps the most widely known is the
Hope Diamond. The stone itself is estimated to be between
1 billion and 3 billion years old. Its extraordinary blue color is
caused by minute traces of boron. Miners extracted the Hope
Diamond from India's Golconda mines during the seventeenth
century. It weighed 112 carats when it was purchased by a
French gem merchant sometime in or around the 1660s; he
subsequently sold it to Louis XIV of France. While in the king's
possession, it was recut to a weight of 67 carats. Later, it was
among the crown jewels that were looted from the French
royal treasury during the French Revolution (1789–1799) and
didn't resurface for many years. Its history remained murky,
likely due to its illicit past, but by 1839 it had ended up in the
hands of the wealthy Hope banking family, from whom it gets
its name. By then, it had been recut again to about 45 carats.
After a storied history among royals, thieves, and merchants
operating both above and below board, the Hope Diamond
rests today in the Smithsonian's permanent collection.

The Koh-i-Noor is a large colorless diamond and among
the largest in the world at approximately 105 carats. It
is currently part of the British crown jewels, but its true
ownership is highly contested; originally mined in India,
it has become a symbol of Britain's colonization of the
country, who has repeatedly asked for it to be returned.

The Daria-i-Noor, which is approximately 185 carats,
currently belongs to the crown jewels of Iran. It is the world's
largest pink diamond and may have been cut down from a
much larger stone. Pink diamonds are among the rarest in the
world; their color is thought to come from the force of colliding
tectonic plates causing deformation in their crystal lattice.

GRAPHITE

Class: Native elements / Hardness: 1–2

Graphite and diamond have identical chemical compositions (both are made of carbon) but very different crystalline structures, so while diamond is a 10 on the Mohs hardness scale, graphite can be as soft as a 1. It is most often found in metamorphic rocks, and its low hardness and dark streak make it the perfect material, mixed with a binding agent, to fill pencils.

SULFUR

Class: Native elements / Hardness: 1.5–2.5

Sulfur, which often forms as crusty masses near volcanos and hot springs, can form well-defined, bright yellow crystals. Sulfur burns easily and gives off a bright blue flame, along with poisonous gas; one of its other names is "brimstone," and it's also called "devil's gold." Sulfur is a common ingredient in everything from winemaking to detergent, but almost always comes from recapturing as a by-product of oil and gas refining. Still, sulfur mining from active volcanoes—which burn blue at night—still takes place in Indonesia, where workers face dangerous health effects.

SILVER

Class: Native elements /
Hardness: 2.5–3.25

Silver, unlike gold, is rarely found in its native state and instead must be extracted from other minerals. It is possible for silver, rarely, to grow as crystals; it is also possible for it to form in a dendritic branching habit, as twisty wires, or in scales. Like gold, it was used to make early coins in many different cultures. Silver is highly conductive of both heat and electricity and so is sought-after in manufacturing.

COPPER

Class: Native elements /
Hardness: 2.5–3

Native copper can form crystals but more frequently is found with a dendritic or filiform habit. Because it can be found in a pure, native state, it was one of the earliest metals used by humans to create beads, knives, and even tools cast in molds. Alloyed with tin, copper became bronze, a harder and corrosion-resistant metal that allowed metallurgy to blossom. Copper can also be added to silver to create sterling silver, a stronger material better suited for jewelry than pure silver.

PLATINUM

Class: Native elements / Hardness: 4–4.5

Platinum is one of the rarest precious metals. It is used in fine jewelry but is also valuable for its industrial applications. Rarely, it can be found in placer form, like gold (see page 68); also, rarely, it can form crystals.

GOLD

Class: Native elements / Hardness: 2.5–3

Gold has been treasured by people throughout history. It is usually found in its native form, either as veins within igneous rock or, more rarely, as "placers"—grains mixed with sand deposited by rivers or glaciers. Gold is usually mixed—alloyed—with other metals when used in jewelry. Karat, not to be confused with carat, is the proportion of gold present in a gold alloy. One gold karat represents 1/24 of a whole; therefore 24 karats is 100 percent gold.

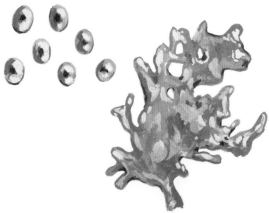

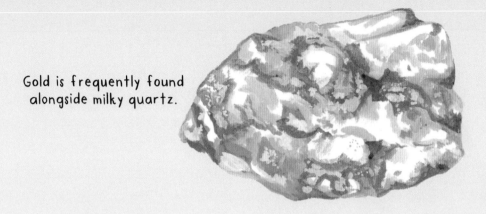

Gold is frequently found alongside milky quartz.

Gold was easily formed into early currency, like this gold Roman Aureus coin.

The Akan peoples of West Africa created intricate and often figurative weights out of brass to measure gold dust for trade.

QUARTZ
Class: Silicates / Hardness: 7

Quartz is a member of the silicate class of minerals, which make up the vast majority of the Earth's crust and upper mantle. Quartz is hard and strong. Varieties include the transparent and colorless rock crystal, the more opaque milky quartz; the pinkish rose quartz; and the warm gray smoky quartz. Plentiful in the Earth, quartz is also plentiful in gem shops, where it is often said to ward off evil and amplify energies.

QUARTZ: Citrine
Class: Silicates / Hardness: 7

Citrine is a yellow-orange or brownish variety of quartz. Its color comes from iron oxide. Oftentimes citrine, also called "burned amethyst," actually is amethyst that has been artificially heat-treated to alter its color; artificially treated citrine is typically browner and naturally occurring citrine, yellower. The same "burning" can take place naturally when amethyst is subjected to high heat from hot magma. Ametrine is quartz with banding of both purple amethyst and yellow citrine.

QUARTZ: Amethyst

Class: Silicates / Hardness: 7

Amethyst's name comes from *amethystos*, which in ancient Greece meant "not intoxicated"; it was once believed that wine imbibed from a vessel of amethyst, or containing amethyst, would not cause drunkenness. Amethyst-containing decanters are still on the market today, though, of course, amethyst does not actually alter the potency of alcohol. Amethyst's characteristic purple color comes from iron.

Amethyst geodes large enough to stand inside of are mined in South America.

QUARTZ: Chalcedony
Class: Silicates / Hardness: 6.5–7

Chalcedony is quartz that is microcrystalline, meaning its crystals are so small that they cannot be distinguished by the naked eye (some samples are cryptocrystalline, with crystals difficult to see even under magnification). Pure chalcedony is white, but the presence of other minerals can impart many colors to chalcedony's varieties.

QUARTZ: Carnelian
Class: Silicates / Hardness: 7

Carnelian is a type of chalcedony. Its orange-to-red coloration is the result of the presence of iron oxide. Its name comes from *carne*, the Latin word for "flesh." It is sometimes called "pigeon blood agate" and is also one of several minerals referred to as "bloodstone." In addition to supposedly treating a long list of blood-related illnesses, carnelian was used as a token to strengthen the voice and become more bold.

QUARTZ: Onyx

Class: Silicates / Hardness: 7

Onyx is a black-and-white–banded variety of chalcedony. Because of its pronounced banding, it has long been a popular choice for cutting cameos, which rely on the stone's contrasting colors to create images. Sardonyx, a related reddish variety, is also especially popular.

The twelfth-century German nun Hildegard of Bingen interwove her spiritual beliefs with a matching conviction in the healing power of gemstones. Her suggested remedy for pain "in the heart or side" instructed the sufferer to hold warmed onyx over a beaker of heated wine until sweat dripped into the wine, creating a potion to be drunk "without delay."

An onyx cameo
viewed from above

An onyx cameo viewed from the side:
the white layer has been carved away
to reveal the black layer underneath

Fire agate

Moss agate

Fortification
agate

QUARTZ: Agate
Class: Silicates / Hardness: 6–7

Agates are chalcedonies that exhibit a number of beautiful appearances. Varieties include fortification agate, which features many concentric bands surrounding one another; fire agate, with its bubbly iridescent blobs; moss agate, which contains beautiful mineral inclusions, reminiscent of moss or seaweed; iris agate, which has extremely thin layers of chalcedony that produce a rainbow of colors when illuminated; grape agate, made of round purple clusters; and lace agate, in which multiple fortification patterns create detailed designs.

Agates generally have been associated with a host of beneficial effects. Agates taken to bed were once thought to improve sleep and bring about good dreams; water that had been used to wash an agate was once believed to promote women's fertility; and farmers wore agate charms to ensure a good harvest. With so many different varieties and endless potential for intricate patterning, it's easy to see why agates have long been a favorite stone.

Iris agate

Grape agate

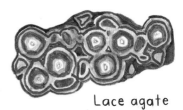

Lace agate

QUARTZ: Opal
Class: Silicates / Hardness: 5–6.5

Opal is a hydrated silica lacking a crystalline structure; technically it is considered a "mineraloid," rather than a true mineral. Its name is believed to come from the Sanskrit *upala*, meaning "precious stone." Opals are divided into three subcategories. Those in the first category, "precious" opals, have a distinctive rainbow iridescence many traditionally associate with the stone, referred to as "showing play of color." The second category, "common" opals, can be brownish orange, coral pink, blue, green, transparent, or colorless and do not show play of color. Finally, the last category, "fire" opals, are bright yellow, orange, or red, and may or may not show play of color. Hyalite, a clear, colorless, and often botryoidal opal, can fluoresce under UV light due to minor uranium impurities.

In 1987, an opal miner discovered a completely opalized fossil of a pliosaur in Australia. The original fossilized skeleton had become a "pseudomorph," a mineral structure in which the original material gets completely replaced by another substance, in this case opal.

An opalized
pliosaur fossil

An opalized
clam fossil

BERYL

Class: Silicates / Hardness: 7.5–8

Beryl is completely colorless when pure. Colorless beryl is sometimes called "goshenite" and was used as lens material for eyeglasses during the Middle Ages. Today, it is sometimes used as a diamond alternative. Beryl is a primary source of beryllium, a material with industrial and scientific applications. It can also be synthesized for use in jewelry.

BERYL: Emerald

Class: Silicates / Hardness: 7.5–8

Aristotle's student Theophrastus believed emeralds had magical powers. They were said to be capable of "resting the eyes." This myth still persists today, as certain beauty products "infused" with emeralds allege that they can brighten the area around the eye.

When heat-treated at high temperatures, greenish beryl can be turned into blue aquamarine.

BERYL: Aquamarine
Class: Silicates / Hardness: 7.5–8

The largest cut aquamarine specimen in the world is the Dom Pedro Aquamarine, named after nineteenth-century Brazilian emperors Pedro I and Pedro II. The crystal from which it was cut was originally three feet (1 meter) long and weighed around one hundred pounds (45 kilograms) before it was dropped by the prospectors who found it. It broke into three pieces, the largest of which was artistically faceted into the Dom Pedro, which now measures over ten thousand carats and is in the Smithsonian museum's collection.

BERYL: Morganite
Class: Silicates / Hardness: 7.5–8

In addition to green emerald and blue aquamarine, beryl can range from colorless to red, pink, or yellow. Pink beryl—a recently popular engagement ring diamond alternative—is also called "morganite," named rather unromantically after American banker J.P. Morgan, an avid collector of precious gems.

BERYL: Heliodor
Class: Silicates / Hardness: 7.5–8

Golden beryl is known as "heliodor," from the Greek *helios*, meaning "sun." One entirely flawless cut heliodor jewel in the Smithsonian museum's collection weighs over two thousand carats.

SODALITE

Class: Silicates / Hardness: 5.5–6

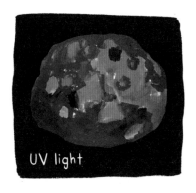

UV light

Sodalite's deep blues and mottled surface—a result of calcite veining—give it an appearance somewhat similar to lapis lazuli. It is often formed into cabochons. Under UV light, sodalite can fluoresce in a variety of colors.

TOURMALINE

Class: Silicates / Hardness: 7–7.5

Tourmaline can take on a wide range of colors, from the electric blue of the Paraiba variety to the opaque black of Schorl tourmaline. But its most distinctive variety is the so-called "watermelon tourmaline," which is "particolored," meaning it features two colors in the same crystal. When sliced crosswise, watermelon tourmaline has a pink center surrounded by a green outer layer. If the specimen is large enough, it can even be cut into a half-pink, half-green gemstone.

JADEITE AND NEPHRITE

Class: Silicates / Hardness: 6–7

Until the nineteenth century, jadeite and nephrite were considered to be the same material—jade. Both stones have been prized material, especially in Eastern cultures, for thousands of years. Statues, weapons, musical instruments, and medicines for kidney ailments were all made from these two types of jade, which were believed to provide their user with a connection to the gods. One Han dynasty prince had a full-body death shroud made entirely of jade tiles. Early Mesoamerican cultures used jade as well; Olmec jade sculptures were created as early as 1000 BCE.

PERIDOT

Class: Silicates / Hardness: 6.5–7

Peridot is the transparent, gem-quality variety of the mineral olivine. It ranges from a pale yellow green to a deep rich green. It's sometimes called "evening emerald," possibly for its tendency to appear a deeper and less yellowy green at night.

TOPAZ

Class: Silicates / Hardness: 8

Topaz can be transparent, but also occurs in a rainbow of colors, including yellow, pink, green, orange, and brown. Topaz can occur in gigantic crystals—some exceptional specimens weigh over one hundred pounds (45 kilograms). Huge topaz crystals were once safe from being cut down into gemstones, since their pale color wasn't particularly desirable; today, topaz can be treated with heat and irradiation to turn it deep blue. The American Golden Topaz is among the world's largest faceted gemstones, weighing over 22,000 carats (more than ten pounds [4.5 kilograms]).

ZIRCON

Class: Silicates / Hardness: 6.5–7.5

Grains of zircon have been dated to 4.2–4.3 billion years old—nearly as old as our Earth itself (which is currently believed to be around 4.54 billion years old), and the oldest minerals identified on Earth. Not to be confused with the similarly named cubic zirconia, zircon is also used as a diamond substitute; naturally occurring zircon is typically brown, but can turn blue or clear via heat treatment.

TANZANITE

Class: Silicates / Hardness: 6.5

Tanzanite is the more well-known name of the purple variety of zoisite, a calcium aluminosilicate. It is generally cut as a gem and can be pleochroic, appearing red, blue, or purple, depending on the angle from which it is viewed. Tanzanite is named after its sole country of origin, Tanzania.

GARNET

Class: Silicates / Hardness: 7–7.5

The name garnet is believed to come from the Latin *granatum*, or "grain," a reference to garnet's resemblance to pomegranate seeds. Red is most associated with the color of garnets, but other colors are possible too, including green or yellow. Garnet is somewhat of an umbrella term for several related gemstones, such as almandite, pyrope, spessartite, and grossularite, among others. Because of their deep red color, garnets are associated with blood. Many purported powers have been bestowed on them over the centuries: the ability to improve blood circulation, relieve menstrual discomfort, or, conversely, cause pain (it was believed that wounds inflicted by garnet bullets could be particularly bloody and slow to heal).

Red garnet cabochons have possibly the goofiest nickname of any gem: "carbuncle."

FELDSPAR

Class: Silicates / Hardness 6–6.5

An abundant silicate, feldspar makes up more than half of the Earth's crust and is a main ingredient of rocks like granite, gneiss, and gabbro.

FELDSPAR: ## Labradorite

Class: Silicates / Hardness: 6–6.5

Gem-quality labradorite is a type of feldspar with a special iridescence called "schiller."

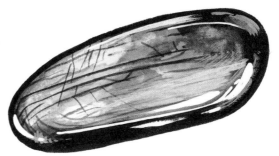

FELDSPAR: ## Amazonite
Class: Silicates / Hardness: 6–6.5

Amazonite is the blue-green, gem-quality variety of the feldspar mineral microcline. It is often carved into cabochons and beads.

FELDSPAR: Moonstone
Class: Silicates / Hardness: 6–6.5

A type of feldspar, moonstone was once believed to be made of actual solidified moonlight. Moonstone is prized for its "adularescence," which is an optical effect created by the gem's layered internal structure. This effect causes ghostly light to appear to float just below the surface of the stone, moving as the stone is turned.

FELDSPAR: Sunstone
Class: Silicates / Hardness: 6–6.5

A number of different feldspars with copper or iron oxide inclusions sparkle like sunlight and are aptly named *sunstone*. This glitter-like effect is referred to as "aventuresence."

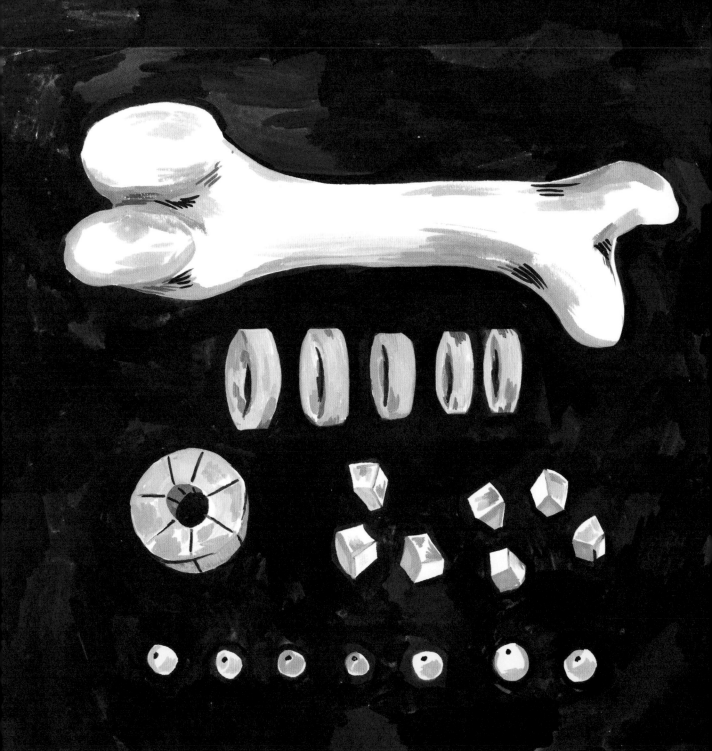

ORGANIC GEMS

Organic gems, including jet, coral, pearls, amber, ivory, and tortoiseshell, are formed by living things—plants, animals, and their by-products—sometimes transformed over time. They have been used much like minerals for as long or even longer in human history, and are often grouped together with true minerals despite their origins in living things—many are as prized as their inorganic counterparts. They carry with them their own special ethical concerns, as the harvesting of many organic gems has consequences for the life-forms and ecologies from which they originate.

JET

Jet comes from the fossilized wood of the monkey puzzle tree. It is formed much like coal—also made from compacted plant remains—but in the sea, rather than in fresh water, like most coal. It is flammable, like coal, and can be intricately carved. The term *jet black*, meaning deep black in color, comes from this stone, whose deep black color is resistant to fading. Whitby, in Yorkshire, England, is the best-known source of jet. Jet has been used since Neolithic times, but became very popular during the Victorian era, particularly after Queen Victoria wore it to mourn the passing of her husband.

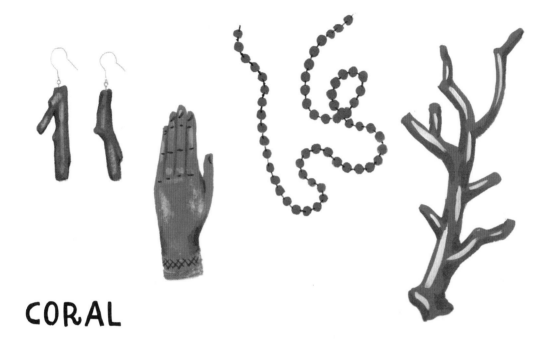

CORAL

Precious coral refers to the slow-growing species of the genus *Corallium*. The portion of the coral that is harvested for human use is the skeleton of the coral created by the coral polyps, and in Corallium can be a vivid pink or bright red. This stony material has been used decoratively for thousands of years.

Some jewelers continue to use harvested coral, taking the stance that Corallium is not substantially threatened the way that shallow-water reef coral is. Other jewelers continue to work only with previously harvested antique coral specimens. The National Oceanic and Atmospheric Administration (NOAA) notes that the United States is currently the largest documented consumer of Corallium species, and recommends avoiding purchasing coral of any kind in order to protect coral populations.

PEARL AND MOTHER OF PEARL

In India, it was once believed that pearls were formed by dewdrops falling into the open mouths of oysters coming to the ocean's surface. In ancient Greece, pearls were thought to be the teardrops of the gods, or, slightly less fancifully though still incorrect, the result of lightning striking shellfish. Pearls actually form as a mollusk's defense mechanism against foreign bodies that penetrate its shell. When predatory intruders—for example, sponges, crabs, or even small fish—pierce the exterior of a mollusk, it responds by covering these intruding bodies in the same substance that coats the inside of its shell: nacre, also known as "mother of pearl." Layers on layers of nacre turn the uncomfortable foreign body into a smooth blob that causes the mollusk less discomfort.

Pearls form in all kinds of shapes, sizes, and colors. Most perfectly spherical pearls are cultured by humans, who intentionally place irritants inside of mollusks. Pearls connected to the interior of a shell are called "blister pearls," and those that float freely within the shell are called "free pearls." Both are prized for jewelry. All shelled mollusks are capable of creating pearls, but only some species create lustrous, gem-quality pearls.

Nacre taken directly from the inside of shells from species like abalone can be worked into jewelry and fine objects, even if it never formed a pearl.

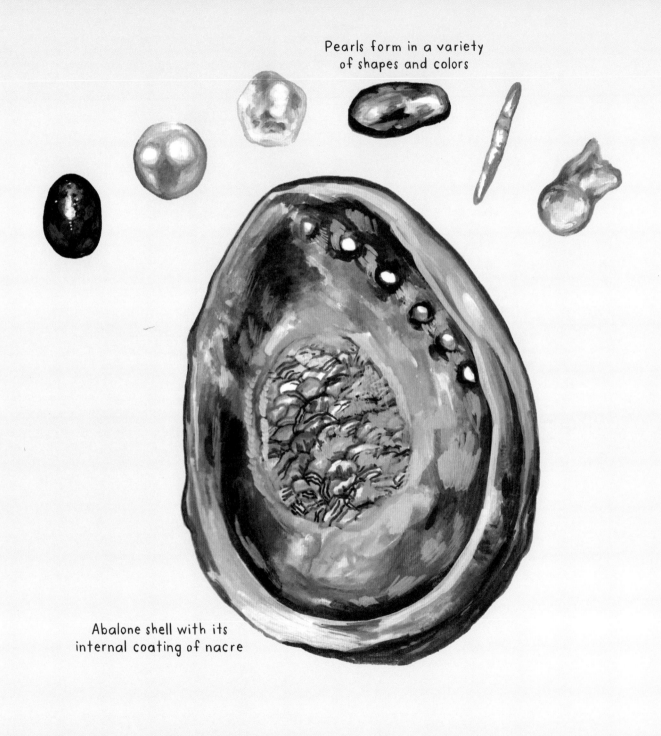

Pearls form in a variety
of shapes and colors

Abalone shell with its
internal coating of nacre

AMBER

Amber is fossilized tree resin. This golden-colored stone, which seems almost to glow when held up to a light source, has often been associated with the sun. True amber is many millions of years old and can house the remains of extinct organisms that were trapped inside its resin at the time of formation. In its use as a gemstone, it is sometimes faceted but more often polished or made into beads or cabochons.

Copal is a younger product of tree resin, which is only semifossilized. It can be used to imitate amber in jewelry and is also burned as traditional incense.

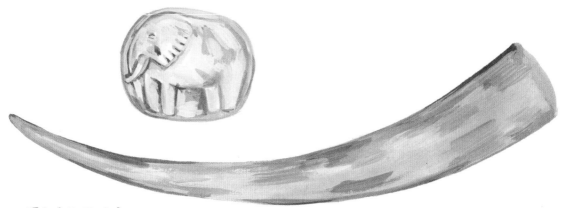

IVORY

Ivory is the material that composes the tusks of elephants. The term also applies to the teeth of other large mammals, including walruses, hippopotamus, narwhals, and sperm whales. Because of the large size of elephant tusks, historically, elephant ivory has been the most valuable. It was used to create piano keys, chess sets, carved sculptures, accessories, and fine art. Over time, demand for elephant ivory led to the dwindling of African elephant populations. Wildlife preservation efforts have been aimed at the elimination of the trade in endangered elephant ivory, and some elephant populations are recovering; others are still in decline. Interest in eliminating ivory use has led to a growing popularity of another organic gem—"vegetable ivory," or the meat of the tagua nut, which can be carved in a similar way to true ivory.

Tagua nut

TORTOISESHELL

Tortoiseshell is made of the scutes—bony plates—that comprise the shell of the critically endangered hawksbill sea turtle. Hawksbill scutes have a mottled brown, black, and golden patterning and could be molded and formed much like plastic, making them an attractive material for pieces such as jewelry boxes, trays, eyeglass frames, and combs for centuries.

Fortunately, tortoiseshell is easily imitated with modern plastics, and a worldwide ban on the hawksbill tortoiseshell trade was put in place in the 1970s. Wildlife conservationists continue to work toward the elimination of the sadly continued illegal trade in hawksbill shells.

ROCKS

As much as we are fascinated with crystals, the humbler and less flashy rock has nevertheless been a part of human culture almost since human culture began. We've even named an important epoch of humanity after stone—the Stone Age, which is the period of time from as early as 3.3 million years ago, when prehistoric humans made the very first stone tools, to around 3000 BCE, when metalworking started to become a widespread practice. That's a pretty large chunk of human history, and rocks have continued to connect us to our ancestors since then. Tools and art made of rock last through the millennia, bearing stories of our own history to us: the enigmatic Stonehenge in England, which shows us that prehistoric humans aligned their rituals with the solstices; the colossal Olmec stone heads from Mexico, carved from basalt, with meanings lost to time; the Rosetta Stone from Egypt, the key that unlocked ancient Egyptian writing systems; and the Korean dolmans, the many thousands of stone grave-marking structures providing remarkably dense evidence of and information about Neolithic people and their culture.

Of course, we are much more likely to encounter endless rocks and very few crystals. And while some are more rare or colorful than others, rocks don't typically have the sparkle or shine of their splashy crystal cousins. But this is one of the great appeals of rocks: they are available. Time spent holding a special stone, turning it over in one's hand, imagining the journey it has taken on a massive time scale to become the object it is today—this can be as special as interacting with a crystal. There is no reason a rock collection should be any less spiritually powerful than a crystal collection; indeed, since rocks are more likely to be collected by one's own hand, rather than being mined from a distant location in potentially dubious working conditions, it could be argued that rock collecting is an ethically preferable practice and possibly even a more spiritually satisfying one. Many of us collected rocks as children. As adults, why not return to this practice? The closer we look at rocks, the more we can see what makes them so appealing.

Rocks are solid aggregates of one or more minerals. They are grouped into three main categories: igneous, sedimentary, and metamorphic. Virtually all rocks begin their lives as igneous rocks, which are formed when magma cools and solidifies. Igneous rocks can be intrusive (formed within the Earth's crust) or extrusive (formed on the surface of the crust, such as by a volcano or on the seabed). Sedimentary rocks are accumulations of particles from eroded rocks or, occasionally, organic materials like seashells. They can be categorized based on the size of the grains that make up the sediments they are formed from: the smallest sediments form mudrocks like shale, while slightly larger grains form rocks like sandstone, and the largest grains form conglomerates. Metamorphic rocks are formed when heat or pressure cause physical or chemical changes to an existing igneous or sedimentary rock.

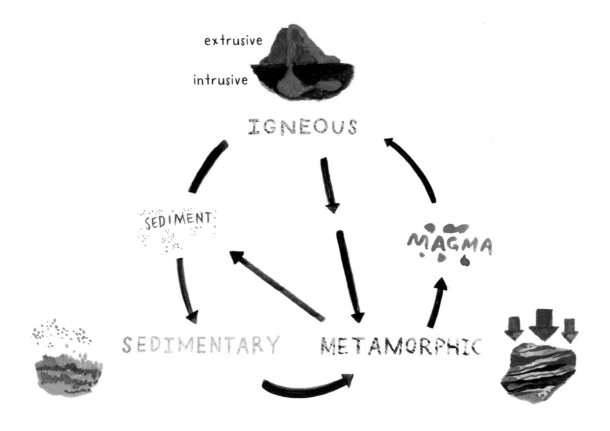

extrusive

intrusive

IGNEOUS

SEDIMENT

MAGMA

SEDIMENTARY

METAMORPHIC

Any rock's life story can be mapped out on the "rock cycle," which demonstrates how interconnected the different rock types can be. An igneous rock can morph into sediment due to weathering over time; this sediment can then form sedimentary rock, which can later be transformed into metamorphic rock through heat and pressure. This rock could even melt back into magma and re-form as a new igneous rock. The cycle isn't a strict, one-direction process, either—igneous rocks can become metamorphic rocks, for instance, and metamorphic rocks can become sedimented; the cycle is more like an interconnected web.

BASALT

Type: Igneous

Basalt is the most abundant igneous rock on the surface of the Earth. It is composed of feldspar, pyroxene, and olivine. A UNESCO World Heritage Site in Ireland with the fable-inspired name the "Giant's Causeway" (right) is made of forty thousand basalt pillars. Although these pillars were naturally formed by the Earth, they appear to have been sculpted by some mythical being.

TUFF

Type: Igneous

Tuff is an umbrella term for porous rock formed from volcanic ash. The Mo'ai heads at Easter Island were carved from tuff, using basalt chisels, and many of the figures were carved directly into the rock face itself before they were cut away and "walked" to their destinations.

Easter Island's
Mo'ai figures

PORPHYRY

Type: Igneous

Porphyry refers to several different configurations of igneous rocks that are mixed with visible crystals. Byzantine royalty placed a high value on this purple rock, believing that those born within the porphyry-lined walls of an imperial palace room were "born in porphyry" and therefore granted an extra dose of royal status. The phrase "*born in the purple*," or born of royal blood, is a reference to this tradition. The power of porphyry was present even in death—at the end of their lives, Byzantine royals were laid to rest in porphyry tombs.

LIMESTONE
Type: Sedimentary

The colossal Great Sphinx of Giza (below), measuring 240 feet (73 meters) long and 66 feet (20 meters) tall, was carved sometime around the twenty-sixth century BCE from a single gigantic piece of limestone that sat as part of the bedrock near the Great Pyramids. Plastered and painted in bright colors, the statue has weathered over the millennia; buried up to its face in sand, it was eventually excavated, and is a subject of continuous concern for preservationists—the soft stone continues to erode to this day.

CHALK

Type: Sedimentary

Chalk is a type of limestone composed almost entirely out of the remains of tiny, ancient marine organisms. The sidewalk and blackboard chalks commonly used today aren't true chalk. Instead, they are usually made out of gypsum (see page 61).

Due to the stark visual contrast between white chalk and the turf that grows above it, huge images, called "hill figures," can be cut into the land above chalk deposits. Although these hill figures are still made today, at least one known example dates to prehistory: the Uffington White Horse in England, which measures 374 feet (114 meters) in length (below).

SHALE
Type: Sedimentary

Shale accounts for about 70 percent of the sedimentary rocks in the Earth's crust, making it the most abundant sedimentary rock. It is composed of hardened clay, mud, or silt. Like crystals, shale can take on different colors, depending on its chemical composition—from the more common gray shale to red, green, brown, and black. Some shale, like that found in deposits near Keokuk, Iowa, along the Mississippi River is "geodiferous"—home to abundant quantities of geodes. And other shale is "oil shale," a source of kerogen, which is a sort of crude version of oil or natural gas. It is mined as a fossil fuel in some parts of the world.

Geodes can occur in shale

COAL

Type: Sedimentary

Coal is a sedimentary rock with organic origins. Rather than being made from compressed bits of minerals, like other sedimentary rocks, coal is made of dead plant debris, compressed over millions of years, during which its mass is converted bit by bit into carbon. The earthy material called "peat" is an intermediary stage, in which plant matter is partially carbonized—peat is not yet a rock. At a high enough percentage of carbon, peat is considered coal, which is divided into three grades: lignite (just enough carbon to be considered coal), then bituminous coal, and finally into the oldest of all, anthracite coal, whose carbon content can reach 95 percent. Because the plant matter at the start of a coal deposit's lifetime may vary greatly in composition—barks, leaves, different types of plants in different environments—the lower grades of coal can differ in quality based on their sources. By the time coal is converted to anthracite, it varies from its fellow anthracites very little, since at that point it is nearly entirely converted to carbon.

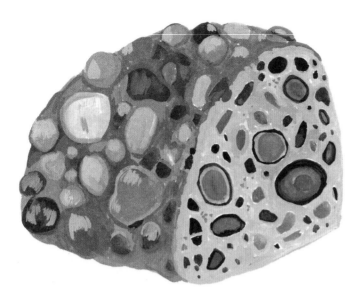

CONGLOMERATE

Type: Sedimentary

Conglomerates are naturally cemented groupings of rocks and minerals that are over 2 millimeters in diameter. One such conglomerate is "puddingstone," which is composed mostly of flint pebbles and bears a striking resemblance to British plum pudding.

LAPIS LAZULI

Type: Metamorphic

Blue is a rare color in nature. Lapis lazuli was thus all the more treasured by ancient civilizations, since only one source of the precious blue stone was known: a mine in a remote part of what is now Afghanistan. Lapis lazuli was of particular importance to the ancient Mesopotamians—along with gold, carnelian, agate, and other precious materials, lapis lazuli adorned the jewelry and headdresses of the Sumerian Queen Puabi and her attendants. These 4,500-year-old artifacts were excavated from the Royal Cemetery of Ur during the twentieth century.

Lapis lazuli is the original source of the prized blue pigment ultramarine. Since natural sources of blue have always been rare and thus conferred high status, blue was the first color people synthesized. Ancient Egyptians created the first synthetic pigment, now referred to as "Egyptian blue," to meet the wide demand for bright blues. Thanks to these synthetic pigments, blues are widely available today.

SLATE

Type: Metamorphic

Slate forms when sedimentary rocks are under specific pressure and temperature conditions. Slate is known for its ability to split into smooth, even sheets. It has long been used as a building material, especially near natural slate supplies, such as those found in Norway, and is still quarried today for use in roofing and landscaping.

MARBLE

Type: Metamorphic

Marble's protoliths are limestone and dolomite. Marble has been a prized material for carving sculptures since as early as the sixth century BCE—possibly even earlier—because of its relative softness.

Glass sponges are
deep-ocean creatures
that produce silica
skeletons that remain
after they die.

MORE MINERAL MARVELS

Our world is rife with mineral marvels. Classified neither as true minerals nor as organic gems, the following are still more mineral oddities that bear mentioning. Many of these are glasses created (or transmuted) by natural events. Also included are accidental gems formed purely as by-products of human activity.

FULGURITE

Fulgurite, also called "lightning glass" or "fossilized lightning," forms when lightning strikes sand or rock. Fulgurite often appears as solid glass covered in partially melted sand.

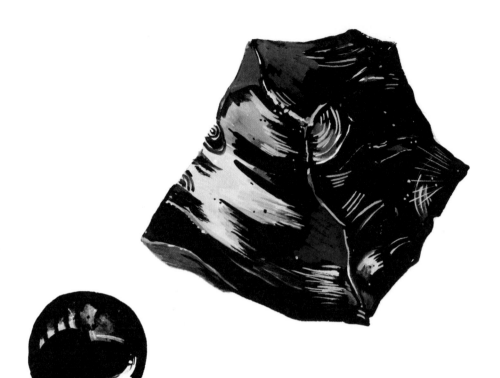

OBSIDIAN

Obsidian is natural volcanic glass that forms when molten lava cools so quickly that it solidifies before crystals have time to grow. "Snowflake" obsidian is named after its characteristic patterning of white cristobalite.

PELE'S HAIR AND PELE'S TEARS

When gas bubbles burst during volcanic eruptions, they can cause molten lava to stretch into thin strings, forming mats of long glassy strands, known as "Pele's hair." Sometimes birds use Pele's hair in their nests. Pele's tears are tiny, hardened droplets of volcanic glass. Both are named after Pele, the Hawaiian goddess of volcanoes.

SEA GLASS

Chunks of manufactured glass, like bottles and jars, can find their way into large bodies of water. There, they tumble and roll in the waves, their sharp edges softening and their surfaces gradually frosting over the years. Sea glass is often collected and made into jewelry, and certain colors are rarer and more desirable than others. In a way, sea glass is a gem made by both humans and the Earth.

SYNTHETIC AND IMITATION GEMS

As modern technology has advanced, humans have managed to re-create geological conditions in laboratories, making it possible to form many types of synthetic gems with machines. Synthetic gems have the same chemical makeup and crystalline structure as their naturally occurring counterparts—for instance, a synthetic diamond and a natural diamond are both "real" diamonds, even though the former can be made in a few weeks in a lab, while the latter formed mysteriously in the depths of the Earth billions of years ago. Synthetic gems can be used in jewelry or for industrial applications that require certain minerals, such as diamond-tipped implements for precision drilling. The production of synthetic gems bears the pronounced benefit of not requiring mining, a process that has historically been fraught with human rights violations.

Imitation gems, unlike synthetic gems, are not made of the same materials as their real counterparts; they only resemble them superficially. They can be made of glass or plastic. Another type of imitation gem is any lower-value gemstone used to mimic an expensive one, such as cubic zirconia, which is a synthetic zirconium dioxide used as an imitation diamond.

The most important aspect of any imitation or synthetic stone is whether its origins are disclosed. Buyers deserve to know what they're buying. Casual deception can take place even at lower price points, so it's best to know what you're looking for. For instance, goldstone and blue goldstone are not stones at all, and neither is opalite—these are examples of glass, usually sold as tumbled lumps or cut gemstones, that are frequently misrepresented as natural stones.

Goldstone

Blue goldstone

Opalite

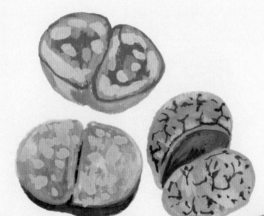

Lithops, or "living stones," are succulent plants that look like pebbles.

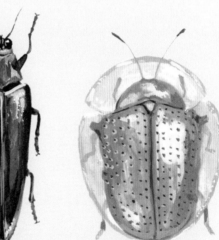

The jewel beetle and golden tortoise beetle are among many species of insect with iridescent and colorful bodies. Jewelry made out of beetle wing coverings does exist!

ACCIDENTAL GEMS

Accidental gems are serendipitous objects resulting from unintentional circumstances. What might normally be overlooked as an inconsequential waste product is instead viewed as a precious, sought-after stone. It's interesting to see what can be achieved by a shift in perspective.

Fordite

Also called "Detroit agate," due to its banded appearance, fordite is the result of automotive paint overspray. Production lines required paint to be sprayed onto cars in such a way that excess paint would collect on the tracks that held the cars. This collected paint layered up over time, hardening into a mass. This mass can be cut and polished just like a stone, which is banded like a fortification agate—both are made of layers upon layers of slightly varying material. Because of advances in automotive production, paint overspray resulting in the formation of fordite is far less likely, making the existing samples all the more precious.

"Wine diamonds," or tartrate crystals, can form in wine because of its naturally high level of tartaric acid.

More as a novelty item than anything, wine diamonds have even been turned into jewelry!

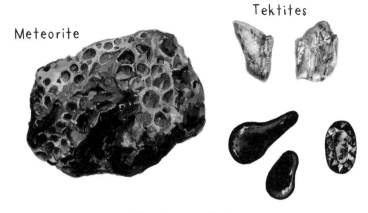

Meteorite

Tektites

METEORITES

Meteorites are solid chunks of debris that have fallen to the surface of a planet from space. They can originate from asteroids, other planets, moons, or even comets. They are typically composed of rock but can also be made mostly or completely of metal. Most meteorites that fall to Earth come from the asteroid belt that lies between Mars and Jupiter. Having formed during the birth of our solar system, many meteorites can be as old or older than the 4.54-billion-year-old Earth itself.

When in space, these rocks are called "meteoroids." While passing through the atmosphere (and potentially making a flash viewable from the surface of the Earth), they are called "meteors." It's not until they reach the surface of the planet that they are referred to as "meteorites."

When a meteorite strikes the Earth, the impact can vaporize the surrounding terrestrial rocks, which can then reform as glassy green, brown, or black objects known as "tektites."

Some meteorites originate from our own moon and are referred to as "lunar meteorites."

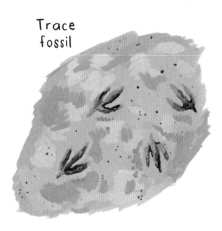
Trace fossil

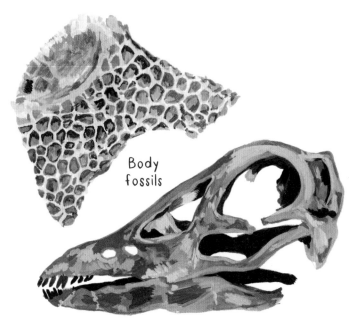
Body fossils

FOSSILS

Fossils are the preserved remains or impressions of living organisms from past geological ages. Fossils can form from animals, plants, or even fungi and bacteria. They are a tremendous source of information about the 3.7-billion-year history of life on our planet.

Trace fossils show evidence of activity from an organism, rather than the organism itself. Examples include footprints, coprolites (fossilized dung), and burrows.

Body fossils are remains of actual body parts or pieces of organisms, like bones and teeth.

Petrified wood is typically formed by chalcedony in a process called "permineralization." The process can be so thorough that sometimes even cell structure is preserved. The presence of other minerals, like iron, carbon, and manganese, can create a rainbow of colors in petrified specimens.

Petrified wood

ACKNOWLEDGMENTS

Thanks are due to Dr. Alex Hastings and to the Science Museum of Minnesota, both of whom generously opened their collections to me. Thanks also to the ever-wonderful team at Ten Speed Press, especially Kaitlin Ketchum, Betsy Stromberg, Kausaur Fahimuddin, Abby Oladipo, Jane Chinn, Chloe Aryeh, and David Hawk. And finally, thanks as always to my loving and supportive family.

ABOUT THE AUTHOR

Kelsey Oseid is a writer and illustrator who also enjoys sewing, knitting, baking, and going on long walks. Kelsey lives in Minnesota with husband Nick, young son Aiden, two cats, and a garden full of flowers, vegetables, and chickens.

INDEX

Published in the United States by Ten Speed Press, an imprint of the Crown Publishing Group, a division of Penguin Random House LLC, New York.
TenSpeed.com

Typeface: FontFont's Suhmo Pro

Library of Congress Cataloging-in-Publication Data
Names: Oseid, Kelsey author
Title: Crystals, rocks, and gemstones: exploring Earth's treasures / Kelsey Oseid.
Description: First edition. | California: Ten Speed Press, [2025] | Includes index.
Identifiers: LCCN 2024039553 (print) | LCCN 2024039554 (ebook) | ISBN 9781984863614 hardcover | ISBN 9781984863621 ebook
Subjects: LCSH: Petrology | Gems | Crystals
Classification: LCC QE432 .O84 2025 (print) | LCC QE432 (ebook) | DDC 552—dc23/eng/20241021
LC record available at https://lccn.loc.gov/2024039553
LC ebook record available at https://lccn.loc.gov/2024039554

ISBN: 978-1-9848-6361-4
EBook ISBN: 978-1-9848-6362-1

Printed in China

Editor: Kaitlin Ketchum
Production editor: Abby Oladipo
Editorial assistant: Kausaur Fahimuddin
Art director and designer: Betsy Stromberg
Production manager: Jane Chinn
Copyeditor: Diana Drew
Proofreaders: Andrea C. Peabbles and Sarah Rutledge
Indexer: Elise Hess
Publicist: David Hawk
Marketer: Chloe Aryeh

10 9 8 7 6 5 4 3 2 1

First Edition

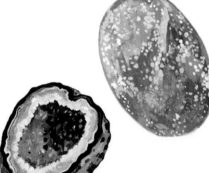

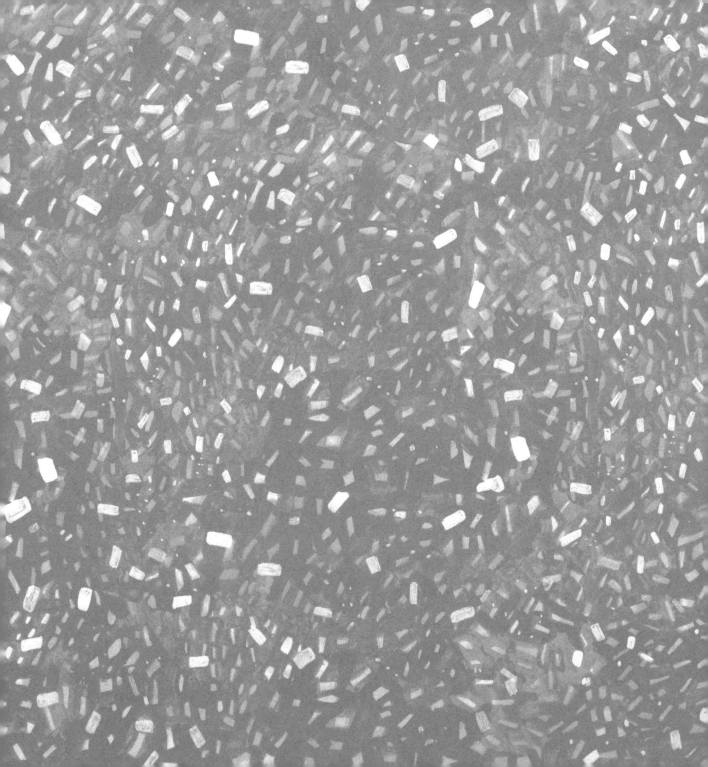